怎麼做蛋最好吃

All about eggs

目錄 Contents

認識蛋 Know Your Eggs

1　蛋黃　Egg Yolk
2　濃蛋白　Inner Thick Egg White
3　稀蛋白　Outer Thin Egg White
4　繫帶　　Chalazae
5　氣室　Air Cell
6　蛋殼　Egg Shell

蛋黃
蛋黃是雞蛋主要的營養來源，其中蛋白質占17.5%、脂肪32.5%、水份48%、礦物質2%和多種微量元素。

濃蛋白
它的功能就好像羊水，具有保護蛋黃的作用。初生蛋的濃蛋白通常會比較多。

稀蛋白
最外圈較稀薄的蛋白。通常老母雞所產的蛋，稀蛋白會比較多。另外，雞蛋放置久了，蛋白所含的黏液素逐漸脫水，也會使蛋白變得稀薄。

繫帶
蛋黃和蛋清之間兩條白色條狀物是繫帶，它可以將蛋黃固定在中央。

氣室
雞蛋較圓的那端有氣室，其中含有空氣。存放雞蛋時，最好將氣室置於上方，避免空氣與蛋黃及蛋白接觸。

蛋殼
主要成份是碳酸鈣，殼上有毛細孔可供呼吸。

Egg Yolk
Egg yolk contains most of the nutrients in an egg. It contains 17.5% protein, 32.5% fat, 48% water, 2% minerals and many types of trace elements.

Inner Thick Egg White
This helps to protect the yolk. Freshly laid eggs have a larger portion of inner thick egg whites.

Outer Thin Egg White
This is the thinner outer layer of the egg white. Eggs laid by older chickens, as well as eggs that have been stored for a longer time, have a larger portion of thin egg whites.

Chalazae
The chalazae are the cordlike strands joining the yolk and the whites, they help to anchor the yolk in the center of the egg.

Air cell
This is a pocket of air at the wide end of an egg. When storing eggs, keep the air cell at the top, this helps to prevent the yolks and whites from coming in contact with air.

Egg Shell
The eggshell is made of calcium carbonate and it has tiny pores on the surface.

蒸蛋 Steamed Eggs

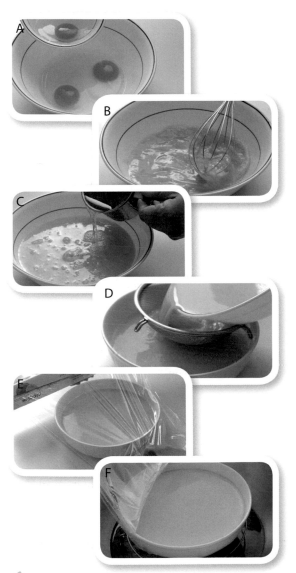

1 將蛋倒入大碗中（圖A）打散，但儘量不要打起泡（圖B）。

2 蒸蛋時要先確定蛋和水的比例，通常1杯蛋汁加2杯水（圖C），熟練後可以加到2杯，會更嫩些。

3 蛋汁用篩子過濾到深的盤子或大碗中（圖D），有泡沫要撈除或用紙巾吸掉。

4 包上保鮮膜（圖E），蒸蛋的時間和所用的容器有關；容器寬、蛋汁淺，蒸的時間就比較短。

5 通常剛開始時可以用大火蒸，蒸至表面已凝固後就要改小火，以免蛋起泡變老。

6 蒸到蛋汁都變為較淺的黃色時，用一支筷子插入蛋中，沒有蛋汁流出，或是晃動蛋碗，可以感覺蛋有彈性，輕微晃動，就是熟了（圖F）

1 Beat eggs in a big bowl (Fig.A). Do not overbeat until foamy (Fig.B).

2 The ratio of eggs to water is 1 cup of beaten egg to 2 cups of water (Fig.C). Once you have mastered the technique, you can increase the amount of water to 2 ½ cups for a softer texture.

3 Strain the egg mixture into a deep dish or bowl (Fig.D), skimming away the foam on the surface with a spoon or paper towel.

4 Cover with cling wrap (Fig.E). The steaming time depends on the plate you use, if you are using a shallow and wide dish, the steaming time is relatively shorter.

5 Use high heat at the beginning, when the surface has set, turn to low heat to prevent bubbles from appearing on the surface and the egg from becoming tough.

6 Once the eggs turn pale yellow, poke a chopstick into it, if the liquid is clear and egg is wobbly but firm, it is cooked (Fig.F).

煎蛋 Fried Eggs

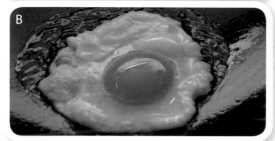

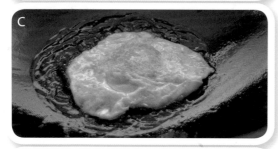

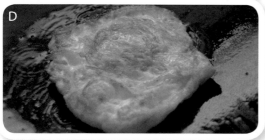

1 鍋燒熱，加入約2大匙油再燒熱，搖盪鍋子，使鍋子均勻沾油，再將油倒出。這個動作稱為"盪鍋"，使鍋子滋潤一下，煎蛋時就不會沾鍋、煎破，如果用不沾鍋就不用這樣處理。

2 另加熱 ½ 大匙油，將一個蛋打入鍋中（圖A），以中火煎約30-40秒鐘，至蛋白凝固（圖B），輕輕翻面，再煎至喜愛的熟度（圖C）。

3 如果喜歡蛋白部分有香酥焦脆的口感，可以用多一些、熱一些的油以大火來煎（圖D）。

1 Heat a pan, add about 2 tbsps of oil and swirl the pan for it to coat evenly with the oil, then pour out the oil. This helps to oil the pan and prevent the egg from sticking onto the pan. However this step is not necessary if you are using a non-stick pan.

2 Heat another ½ tbsp of oil. Break an egg into the pan (Fig.A). Fry over medium heat for about 30-40 seconds until the egg white has set (Fig.B), gently turn over and continue to cook it to your preferred doneness (Fig.C).

3 If you prefer your fried eggs to have a slight crispy edge, use more oil and cook over high heat (Fig.D).

煎蛋皮 Making Crepes

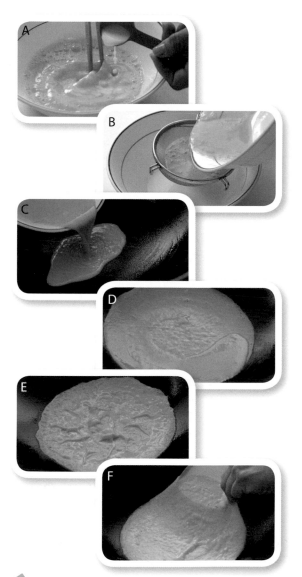

1 要用來包卷材料的蛋皮，記得在蛋液中加入少許調勻的太白粉水，以增加蛋皮的彈性（圖A）。

2 蛋汁打好後最好過濾一次（圖B），且不要有泡沫，可使蛋皮均勻光滑些。

3 鍋燒熱後用紙巾塗抹少許油，油不宜太多，因為會使蛋皮滑動難成形（圖C）。

4 倒下蛋汁的同時就要轉動鍋子（圖D），形成所要的大小，再繼續轉動鍋子，達到所需的厚薄度（圖E）。

5 要再蒸或煎過的蛋皮，只要煎一面熟，另一面只要凝固即可（圖F），直接要吃的就要翻面再煎熟。

1 If the crepes are meant for wrapping other ingredients, add a little potato starch mixture to the egg mixture (Fig.A).

2 Strain the egg mixture (Fig.B). For a smoother texture, make sure there is no bubbles in the mixture.

3 Heat a pan, wipe it with a little oil. If too much oil is used, the crepe will not hold its shape well (Fig.C).

4 Swirl the pan (Fig.D) to spread out the egg mixture to your preferred size and thickness (Fig.E).

5 If you still need to steam or pan-fry the crepes later on, just pan-fry until one side is cooked and the other side has just set (Fig.F). If you are serving the crepes immediately, fry both sides until cooked.

炒蛋 Scrambled Eggs

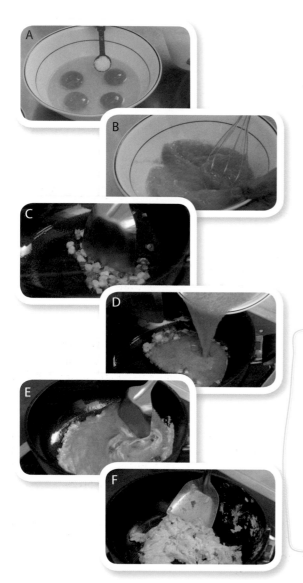

1　蛋打在碗中，加鹽調味（圖A），以打蛋器充分打散（圖B）。

2　起油鍋，爆香蔥花（圖C），倒入蛋汁（圖D）。

3　用鍋鏟快速移動蛋汁（圖E）。

4　當蛋全部凝結熟成即可起鍋（圖F）。

1　Break eggs into a bowl and season with salt (Fig.A). Beat well with a whisk (Fig.B).

2　Heat oil in a pan, fry chopped spring onions until fragrant (Fig. C), then add the eggs (Fig.D).

3　Use a spatula to briskly stir the eggs (Fig.E).

4　Dish out when the eggs are cooked (Fig.F).

NOTE

· 炒蛋時，油要多一些，火要大一些，才能炒出蛋的香氣。

· 也可在蛋汁中加一些水或鮮奶，會使蛋炒得更嫩一些。

· 除蔥花外，許多材料都可以用來搭配炒蛋，例如：韭菜、九層塔、火腿、香腸、培根、蝦仁、洋蔥等。

· Make sure you use enough oil and cook the eggs over high heat for a better fragrance.

· For a more delicate texture, add some water or fresh milk to the egg mixture.

· Besides spring onions, you may also add chives, basil, ham, sausages, bacon, shrimps or onions.

小小雞蛋妙用多多！The Humble Egg and Its Many Uses

小小的一顆雞蛋，除了能食用，用處還真多！神奇的蛋殼除了可以拿來和小朋友們一起做手工，它還是廚房的清潔好幫手呢！

清洗鍋子
取一把壓碎了的蛋殼，加一點水，滴一小滴清潔劑，就可以把骯髒的鍋子洗得乾乾淨淨！

清洗熱水瓶
在熱水瓶裡放入一點粗鹽和一個捏碎了的蛋殼。加一點水，然後用力搖晃，再用清水沖洗就可以了。如果保溫瓶裡有洗不掉的汙漬，也可以用同樣方法，放置隔夜，然後用水沖洗乾淨即可。

清洗果汁機或食物處理器
把兩個壓碎了的蛋殼放入果汁機中，加入500毫升的水一起攪打，就可以輕易洗掉果汁機死角的髒汙。骯髒的食物處理器也可以用同樣方法來清洗。

Eggs aren't just for eating! Besides using egg shells for doing crafts with the kids, they also prove to be very useful around the house espcially in household cleaning! You don't need expensive and harsh cleasing agents and these eggshells are all natural for the environment.

Clean Your Pots and Pans
Mix a handful of crushed eggshells with some water and maybe a drop of detergent to scrub your kitchen pots and pans. You'll be surprised to find that the eggshells can clean off any residue easily.

Clean Vacuum Flask or Thermos
Put some coarse salt and crushed egg shells in a vacuum flask. Pour in some water and shake well, then rinse with water. To clean a stained thermos, put some crushed egg shells in a dampened thermos and let it sit overnight. Add water and rinse well.

Clean Blender or Food Processor
Crush 2 egg shells and place them in the blender, add about 500 ml of water and switch it on. This is useful for getting rid of any stains and dirt at corners where you can't reach by hand.

剝蛋殼有技巧！
What is the best way to peel a hard boiled egg?

剝蛋殼時一不小心就會連蛋白一起剝掉，只要將煮好的水煮蛋沖冷水後泡入冷水中約3-5分鐘，蛋殼熱漲收縮後更容易剝開。敲蛋殼時，要從氣室也就是圓的那端敲開，才不會敲破蛋白。

Immerse cooked hard boiled eggs in cold water for about 3-5 minutes. Then start peeling at the air bubble in the wide end of the egg. The egg shell will then come off easily.

氣室
Air cell

9分鐘
9 minutes

7分鐘
7 minutes

11分鐘
11 minutes

5分鐘
5 minutes

3分鐘
3 minutes

- 冰箱取出的蛋，在煮之前要先放在室溫回溫一下，再放入水中加熱煮熟，這樣就可以避免冷熱溫差過大使蛋殼破裂。
- 若煮的是鴨蛋，必須在水滾後再加入，因為鴨蛋蛋殼較厚，這樣比較容易計時。
- Keep refrigerated eggs at room temperature for a short while before cooking. This prevents the shells from cracking due to a sudden increase in temperature.
- If you are boiling duck eggs, add the eggs only when the water has come to a boil as they have thicker shells, this makes it easier to gauge the actual cooking time.

水煮蛋
boiled eggs

材料
蛋　　　　　數個
鹽1茶匙或醋1大匙

做法
1　冰箱取出的蛋，要先放在室溫回溫。
2　蛋放入冷水鍋中，水要超過蛋的高度約2公分（圖A）。
3　水中加1茶匙鹽（圖B）或1大匙醋，可以讓蛋殼不易破裂。
4　先以大火煮至水滾後，改以小火煮至喜愛的熟度。
5　在剛開始煮時要用筷子轉動蛋（圖C），以使蛋黃能凝固在蛋的中間。

Ingredients
a few eggs
1 tsp salt or 1 tbsp vinegar

Method
1　Keep refrigerated eggs at room temperature for a short while before cooking.
2　Place eggs in cold water. The water level must be higher than the eggs by 2cm (Fig.A).
3　Adding 1 tsp of salt or 1 tbsp of vinegar to the water will help keep the egg whites from running out of any eggs that crack while cooking (Fig.B).
4　Bring to boil over high heat, then continue to cook over low heat to your preferred doneness.
5　Stir the water when you start boiling the eggs to help keep the egg yolks at the center (Fig.C).

A

B

C

蒸蛋時，蛋和水的比例是多少？

What is the correct proportion of egg and water when steaming egg?

蒸蛋的水和蛋比例很重要，調蛋汁時應以1杯蛋汁加2杯水，如果想更嫩些，水的份量可以加到2.5杯。

The ratio of eggs to water is 1 cup of beaten egg to 2 cups of water. For a softer texture, you might want to increase the amount of water to 2.5 cups.

蠔油蛤蜊蛋
steamed eggs with clams

- 購買蛤蜊時，應選擇緊密合著的，若發現殼全開或有裂痕，則表示已經不新鮮了。
- 可以在碗上蓋上一張保鮮膜，蒸好的蛋會很光滑。
- *When buying clams, never select a clam that is chipped or broken and do not choose a clam that is open. Fresh live clams should be tightly closed.*
- *Cover a sheet of cling wrap on the dish for smooth steamed eggs.*

材料
蛋	4個
冷清湯或水	2杯
蛤蜊	15粒
蔥	1根
薑	1片
香菜或紅蔥酥	適量

調味料
鹽	⅓茶匙
酒	1茶匙
蠔油	1大匙
鹽	¼茶匙
胡椒粉	適量
太白粉	適量

做法

1. 蛋加鹽打散，加入清湯調勻，過濾到深盤中，入蒸鍋，以小火蒸熟。

2. 蛤蜊用清水1¼杯煮至殼微開即撈出，剝下肉，湯汁留下備用。

3. 用1大匙油煎香蔥和薑片，淋上酒和蛤蜊湯，煮滾，撈棄蔥和薑。

4. 加蠔油、鹽和胡椒粉調味，勾芡後加入蛤蜊肉，煮滾，全部淋到蛋面上，可撒下少許香菜或紅蔥酥增香。

Ingredients
4 eggs
2 cups stock or water
15 clams
1 stalk spring onion
1 slice ginger
some coriander or fried shallots

Seasonings
⅓ tsp salt
1 tsp wine
1 tbsp oyster sauce
¼ tsp salt
some pepper
some potato starch mixture

Method

1. Beat eggs with salt, add stock and strain into a deep dish. Steam over low heat until cooked.

2. Cook clams in 1 ¼ cups of water. Extract the meat and reserve the liquid.

3. Heat 1 tbsp of oil and saute spring onion and ginger until fragrant, add wine and clams liquid, bring to a boil. Discard spring onion and ginger.

4. Add oyster sauce, salt and pepper, then thicken with potato starch mixture. Add clams and bring to a boil. Pour over steamed eggs. Scatter over coriander or fried shallots to enhance the flavor.

蒸肉餅要注意的事！
What should you take note of when making steamed meat patties?

蒸的時間要視肉餅的薄厚而定，也許要縮短或加長時間；蒸盤要抹上一點油，蒸好的肉才不會黏盤。

The steaming time depends on the thickness of the patties. Check for doneness and adjust the time accordingly. Smear some oil in the steaming dish to prevent meat from sticking onto the dish.

鹹蛋蒸肉餅
steamed pork patties with salted egg

TASTY TIPS

- 鹹蛋黃壓住的地方，絞肉較不易熟，因此要翻面查看，確定鹹蛋下面的絞肉已熟。
- 也可在絞肉裡加入切碎的馬蹄，增加口感。
- The meat underneath the egg yolks will take a longer time to cook, thus you need to turn the yolks over once to check for doneness.
- Add chopped water chestnuts to the minced meat for a crunchy texture.

材料

豬前腿絞肉	250克
大蒜泥	½茶匙
蔥花	2大匙
生鹹鴨蛋	2個

調味料

鹽	¼茶匙
水	2-3大匙
酒	1茶匙
醬油	½大匙
胡椒粉	⅙茶匙
糖	¼茶匙
太白粉	2茶匙

做法

1 將絞肉再剁細一點，放入碗中，加入大蒜泥、蔥花、鹽和水攪拌至有黏性。

2 加入酒、醬油、胡椒粉、糖和1個鹹鴨蛋蛋白，攪拌均勻，最後加入太白粉拌勻，放入一個有深度的盤子裡。

3 鹹鴨蛋取蛋黃，切成兩半，圓面朝下放在絞肉上。

4 放入蒸鍋中，以大火蒸熟，約25分鐘，蒸至最後5分鐘，將鹹蛋黃翻面，再繼續蒸至熟即可。

Ingredients

250g minced arm shoulder pork
½ tsp minced garlic
2 tbsps chopped spring onions
2 raw salted eggs

Seasonings

¼ tsp salt
2-3 tbsps water
1 tsp wine
½ tbsp dark soy sauce
⅙ tsp pepper
¼ tsp sugar
2 tsps potato starch

Method

1 Chop minced pork finely and stir well with minced garlic, spring onions, salt and water until sticky.

2 Add wine, dark soy sauce, pepper, sugar and 1 salted egg white. Mix well, then add potato starch and stir well. Place meat mixture in a deep dish.

3 Cut salted egg yolks into halves, place on the meat, rounded sides down.

4 Steam over high heat for about 25 minutes. At the last 5 minutes, turn over the salted egg yolks and continue to steam until cooked.

如何變化出不一樣的蛋捲？
What are the different ways of serving this dish?

雞肉可以用其他肉類取代，甚至可以使用煮熟的鵪鶉蛋，造型可愛又美味。也可用其他蔬果如黃瓜、蘋果等取代小番茄。

Replace the chicken meat with other types of meat, or use quail eggs instead for an attractive presentation. You may also use other fruits or vegetables such as cucumbers and apples.

蔬果雞肉蛋捲
chicken and egg skewers

雞胸肉去皮，可以減少油脂的攝取量，吃起來更為健康。

- *Make this appetizer dish healthier by removing the skin of the meat to reduce fats intake.*

材料
雞胸肉	2片
蛋	3個
小番茄	10顆

調味料
醬青	適量
五香粉	適量
白糖	適量
胡椒粉	適量

做法
1 雞胸肉洗淨，切成丁狀，加入適量醬青和五香粉醃製約15分鐘。

2 將醃好的雞胸肉放入鍋中，煎至表皮呈金黃色。

3 蛋打散，加入白糖和胡椒粉。

4 將蛋液倒入平底鍋，以小火煎成蛋皮，起鍋，將蛋皮切成長條形。

5 雞肉放在蛋皮上，卷起再用牙籤固定。

6 在頂端加插小番茄。

Ingredients
2 pieces chicken breast meat
3 eggs
10 cherry tomatoes

Seasonings
some light soy sauce
some five spice powder
some sugar
some pepper

Method
1 Wash and diced chicken breast meat. Add some light soy sauce and five spice powder, set aside for 15 minutes.

2 Pan-fry chicken until golden brown.

3 Beat eggs and add sugar and pepper.

4 Pour egg mixture into a flat pan and cook into a crepe using low heat. Remove from the pan and slice into long strips.

5 Roll chicken with egg crepe and secure with a toothpick.

6 Add cherry tomatoes.

如何使煮好的蝦捲起？
How do you prepare prawns so that they curl up into a ball when cooked?

在蝦的背部劃一刀口，炸好的蝦就會捲起呈球狀。
Cut a slit down the back of the prawn, it will curl up into a ball when cooked.

TASTY TIPS

- 剩下的鹹蛋白可以用來炒飯或配粥吃，也很可口。
- 炸油的溫度應該大約190°C，如果油溫太低，蝦和墨魚炸得太久，肉質會變老。
- Use the remaining egg whites to cook fried-rice or serve with porridge.
- The ideal oil temperature for deep-frying seafood is about 190°C, too low and the seafood will take too long to cook thus turning rubbery and tough.

金沙雙鮮
cuttlefish and prawns with golden crumbs

材料

墨魚	300克
新鮮草蝦	500克
玉米粉	½杯
熟鹹鴨蛋	5個
白芝麻	1大匙
咖喱粉	1茶匙
糖	1茶匙

醃料

鹽	¼茶匙
酒	½茶匙
胡椒粉	¼茶匙
蛋白	1大匙
太白粉	½茶匙

做法

1 墨魚剝去外皮後切成5公分長的粗條；草蝦剝殼，尾巴留著，在背部劃一刀口。兩者一起加入醃料醃10分鐘，沾裹玉米粉。

2 鹹蛋取蛋黃，切丁備用。

3 燒熱炸油，分別將墨魚和蝦炸熟，撈出瀝乾油份。

4 另用一個鍋子燒熱油，放入鹹蛋黃炒散成泡沫，取一半與炸好的蝦拌炒，盛出，撒下炒香的白芝麻。

5 另一半鹹蛋黃加咖喱粉和糖炒勻，放入墨魚拌勻，盛出。

Ingredients

300g cuttlefish
500g prawns
½ cup corn flour
5 cooked salted eggs
1 tbsp white sesame seeds
1 tsp curry powder
1 tsp sugar

Marinade

¼ tsp salt
½ tsp wine
¼ tsp pepper
1 tbsp egg white
½ tsp potato starch

Method

1 Peel away the skin from the cuttlefish, then cut into 5cm long thick pieces. Peel the prawn shells, leaving the tails intact and cut a slit down the back. Mix both with marinade and set aside for 10 minutes. Dredge both cuttlefish and prawns with corn flour.

2 Take out the yolks from the salted eggs and cut into dices.

3 Heat oil and deep-fry cuttlefish and prawns. Drain away the oil.

4 Heat oil in another pan and fry salted egg yolks until foamy, use half of the yolks to combine well with the prawns. Dish out prawns and scatter with white sesame seeds.

5 Add curry powder and sugar to the other half of the yolks and stir well with the cuttlefish. Serve.

煮蛋時，如何避免蛋殼因為有裂痕而造成蛋白溢出？

How do you prevent egg whites from coming out of the eggs should they crack while boiling?

要避免蛋有裂痕時蛋白溢出，可以在水中加1茶匙鹽或一大匙醋。

Adding 1 tsp of salt or 1 tbsp of vinegar to the water will help keep the egg whites from running out of any eggs that crack while cooking.

- · 煮好的蛋立刻泡冷水，就會很容易剝殼。
- · 水煮蛋可以在冰箱內保存一個星期，在室溫則最好不要超過兩小時。
- · Immerse boiled eggs in cold water, this makes it very easy to peel away the shells.
- · Hard-boiled eggs can last a week in the refrigerator. However it shouldn't be kept at room temperature for more than two hours.

回鍋蛋
twice-cooked eggs

材料

蛋	4個
紅辣椒	2條
蔥	2根

調味料

豆豉	1茶匙
醬油	1大匙
鹽	¼茶匙
糖	½茶匙
醋	2茶匙
水	2大匙

做法

1 蛋放入鍋中，加冷水，水中加少許鹽或醋，以大火煮滾後改中小火煮約11分鐘至全熟。

2 蛋取出泡冷水至涼，剝殼，再小心切成片，儘量避免蛋黃與蛋白脫離。

3 紅辣椒去籽，切片；蔥切粒。

4 燒熱2大匙油，放下蛋片煎黃，翻面再煎一下，撒下紅辣椒粒爆香，再倒入調勻的調味料煮香，輕輕拌合蛋片，撒下蔥花，略拌勻即可盛出。

Ingredients

4 eggs
2 red chillies
2 stalks spring onion

Seasonings

1 tsp fermented black beans
1 tbsp dark soy sauce
¼ tsp salt
½ tsp sugar
2 tsps vinegar
2 tbsps water

Method

1 Place eggs in a pot with cold water (add a little salt or vinegar to the water), bring to a boil over high heat, then boil over medium low heat for about 11 minutes until they are cooked.

2 Remove eggs and immerse in cold water. Peel the eggs and carefully cut into slices. Try to keep the egg yolks and whites intact.

3 Remove seeds from the red chillies and cut into slices. Chop spring onion.

4 Heat 2 tbsps of oil and pan-fry the egg slices until they are slightly browned on one side, turn over and fry again. Add chillies and fry until fragrant, then add the combined seasonings. Gently stir well with the egg slices, toss in spring onion and serve.

除了莧菜，還有什麼蔬菜也適合用來做這道菜？
What other kinds of vegetables are suitable for making this dish?

軟嫩的綠葉蔬菜如菠菜、豆苗都很適合用來做這道菜。
Soft leafy vegetables such as spinach and pea shoots are all suitable for making this dish.

金銀蛋莧菜
chinese spinach with salted egg and century egg

材料

熟鹹蛋	1個
皮蛋	2個
莧菜	300克
蔥	1根
大蒜末	1茶匙
高湯	1杯

調味料

| 鹽 | 適量 |
| 胡椒粉 | 適量 |

做法

1　鹹蛋切小塊。

2　皮蛋放入水中煮5-6分鐘，沖冷水至涼後切小塊。

3　莧菜摘好，放入熱水中燙一下，撈出，沖涼。

4　燒熱油爆香蔥段和蒜末，放入莧菜和高湯，煮至莧菜微軟，加入皮蛋和鹹蛋同煮2分鐘即可。

Ingredients
1 cooked salted egg
2 century eggs
300g Chinese spinach (yin choi)
1 stalk spring onion
1 tsp chopped garlic
1 cup stock

Seasonings
dash of salt
dash of pepper

Method

1　Cut salted egg into small pieces.

2　Boil century eggs for 5-6 minutes, rinse under cold water before cutting into small pieces.

3　Trim Chinese spinach and blanch briefly in boiling water. Remove and rinse under cold water.

4　Heat oil and saute spring onion and garlic until fragrant. Add Chinese spinach and stock to cook until vegetables are soft, then add century eggs and salted egg to cook for another 2 minutes.

如何使蛋黃煮熟後凝固在中間？

How do you ensure that the egg yolk stays at the center of the egg?

在剛開始煮蛋時，要用筷子在水裡轉動蛋，這樣蛋煮好後蛋黃就能凝固在蛋的中間。

Using a pair of chopsticks, keep stirring the eggs in the water at the initial stage of boiling, this will keep the egg yolk at the center of the egg.

炸百花蛋
deep fried stuffed eggs

TASTY TIPS

- 油炸時，油溫不要太高，以免表面炸焦了，而蝦餡仍未熟透。
- *Make sure that the oil for frying is not too hot, or else the surface gets burnt while the prawn paste remains uncooked.*

材料

蛋	6個
蝦仁	240克
馬蹄	5粒
米粉或芥蘭菜葉	適量

調味料

蛋白	1大匙
鹽	½茶匙
酒	1茶匙
太白粉	½大匙
麻油	適量

做法

1 蛋放入鍋中，加冷水，水中加少許鹽或醋，以大火煮滾後改中小火煮約11分鐘至全熟。蛋取出泡冷水至涼，剝殼，每個蛋對切成兩半，撒少許鹽，擱置片刻。

2 蝦仁壓成泥狀，加入切碎、擠乾水份的馬蹄，加入調味料仔細拌至均勻，成為有彈性的蝦餡。

3 在蛋的切口上撒上少許太白粉，再放上1大匙的蝦餡，手指沾水，抹光表面，使之成整的蛋形，再沾上一層太白粉。

4 燒熱炸油，放下百花蛋，用慢火炸熟蝦面，使之成為金黃色，撈出。

5 每個橫切成兩半，附上炸松的米粉或芥蘭菜葉即可。

Ingredients

6 eggs
240g shrimps
5 pieces water chestnuts
some vermicelli or mustard greens

Seasonings

1 tbsp egg white
½ tsp salt
1 tsp wine
½ tbsp potato starch
few drops of sesame oil

Method

1 Place eggs in a pot with cold water (add a little salt or vinegar to the water), bring to a boil over high heat, then boil over medium low heat for about 11 minutes until they are cooked. Remove eggs and immerse in cold water. Peel the eggs and carefully cut into halves, sprinkle salt on the cooked eggs.

2 Smash the shrimp and combine well with the chopped and squeezed dry water chestnuts and seasonings until sticky.

3 Sprinkle potato starch on the cut surface of the eggs, use 1 tbsp of prawn mixture to form into the shape of an egg with wet fingers, then dredge with potato starch.

4 Heat oil and deep-fry eggs over low heat until the shrimp is cooked and golden brown, drain.

5 Cut each egg in half. Serve with fried vermicelli or mustard green.

美乃滋自己做！
How to make mayonnaise at home?

蛋黃2個慢慢加入1-1杯橄欖油，邊加邊打勻，使蛋黃和橄欖油完全融合，當蛋黃變濃稠時，依個人口味加檸檬汁和鹽、糖、芥末醬等調味，拌打均勻即可。

Gradually beat 2 egg yolks into 1-1 ½ cups of olive oil until they are well combined. When the mixture has thickened, season with lemon juice, salt, sugar or mustard to taste and mix well.

馬鈴薯蛋沙拉
potato egg salad

TASTY TIPS

可隨個人喜愛加入不同的蔬果，如番茄、奇異果、櫻桃等。

- Tomato, kiwi fruit and cherry can also be used.

材料

馬鈴薯	2個
紅蘿蔔	1條
蛋	5個
小黃瓜	1條
蘋果	1個
美奶滋	4-5大匙

調味料

鹽	½茶匙
黑胡椒粉	適量

做法

1　馬鈴薯、紅蘿蔔和蛋洗淨，放入鍋中，加水煮熟，先取出紅蘿蔔，約12分鐘時取出蛋，馬鈴薯煮至軟。

2　將紅蘿蔔切小片；蛋切碎；馬鈴薯削皮，切成塊。

3　黃瓜切片，用少許鹽醃一下，擠乾水份；蘋果連皮切丁。

4　所有材料放入碗中，加入美奶滋和調味料拌勻，放入冰箱1小時即可食用。

Ingredients
2 potatoes
1 carrot
5 eggs
1 Japanese cucumber
1 apple
4-5 tbsps mayonnaise

Seasonings
½ tsp salt
dash of black pepper

Method

1　Rinse potatoes, carrot, eggs and put in a pot, add water to cook. Remove carrot when it is cooked, then take out the eggs after about 12 minutes. Cook potatoes until they are soft.

2　Cut carrot into slices and chop the eggs. Peel potatoes and cut into chunks.

3　Cut cucumber into slices and rub with some salt. Set aside for a while before squeezing out excess liquid. Cut apple into dices, leaving the skin on.

4　Combine all the ingredients with mayonnaise and seasonings. Leave to chill in the refrigerator for an hour before serving.

用來炸的皮蛋為什麼要事先煮過？
Why is it necessary to boil century eggs first prior to deep-frying?

先將皮蛋煮過，可以讓蛋黃凝固，之後再油炸的時候就不易脫落。
Century eggs need to be boiled first to allow the yolks to set before deep-frying so that they hold their shape better when frying.

- 按照同樣做法，還可以做成宮保皮蛋、椒鹽皮蛋。
- 皮蛋要先沾乾麵粉，讓麵粉吸乾表面的水分，之後再沾裹麵糊，在油炸時麵糊就不容易脫落。
- 取一條細線，一端用牙咬住，一端用手拉緊，將皮蛋線上上分切，又快又美觀。

- Use the same method to make century eggs with dried chillies or century eggs with pepper and salt.
- First dredge the eggs with flour, this allows the flour to soak up any liquid so that the batter stays on better when deep-frying.
- To cut century eggs neatly and quickly, use a thread instead of a knife. Take a piece of thread, hold one end in your mouth, pull it taut with your hand and use it to cut the eggs.

雙味皮蛋
spicy century eggs, sweet and sour century eggs

材料

皮蛋	5個
麵粉	2大匙
絞肉	2大匙
大蒜末	1茶匙
紅辣椒末	1茶匙
蔥末	½大匙

麵糊

水、麵粉、玉米粉、糯米粉 各適量

調味料 1

醬油	1½大匙
鹽	⅓茶匙
糖	1茶匙
水	½杯
醋	2茶匙
麻油	數滴
花椒粉	½茶匙

調味料 2

番茄醬	2大匙
糖	3大匙
醋	3大匙
鹽	¼茶匙
水	¼杯
太白粉	1茶匙

做法

1. 皮蛋煮5分鐘，取出泡冷水，剝殼後一切為6片；麵糊材料調勻。

2. 皮蛋先沾乾麵粉後再沾裏麵糊，放入熱油中炸至金黃，撈出，瀝乾油。

3. 燒熱1大匙油，炒散絞肉，再放入蒜末炒香，加入醬油、鹽、糖和水炒合，淋下醋和麻油，熄火。

4. 放入一半的皮蛋，撒下紅辣椒、蔥末和花椒粉拌合，即成麻辣皮蛋，盛出。

5. 小鍋中放入調味料2煮滾，放入另一半的皮蛋略拌合，即成糖醋皮蛋，盛出。

Ingredients

5 century eggs
2 tbsps flour
2 tbsps minced meat
1 tsp chopped garlic
1 tsp chopped red chilli
½ tbsp chopped spring onion

Batter

some water, flour, corn flour, glutinous flour

Seasonings 1

1 ½ tbsps dark soy sauce
⅓ tsp salt
1 tsp sugar
½ cup water
2 tsps vinegar
few drops of sesame oil
½ tsp Sichuan pepper

Seasonings 2

2 tbsps tomato sauce
3 tbsps sugar
3 tbsps vinegar
¼ tsp salt
¼ cup water
1 tsp potato starch

Method

1. Boil century eggs for 5 minutes. Remove and immerse in cold water, then peel and cut each egg into 6 pieces. Combine well the ingredients for the batter.

2. Dredge eggs in flour before coating with the batter. Deep-fry until golden brown. Remove and drain away the oil.

3. To make spicy century eggs: Heat 1 tbsp of oil and fry minced meat until separated. Add garlic and fry until fragrant, stir in dark soy sauce, salt, sugar and water. Drizzle over vinegar and sesame oil. Turn off the heat.

4. Add half of the century eggs and scatter over red chilli, spring onion and Sichuan pepper. Combine well. Dish out.

5. To make sweet and sour century eggs: Bring seasonings 2 to a boil in a pot, add the other half of the century eggs and combine well. Serve.

浸泡粉絲要注意些什麼？
What is the correct way to soak vermicelli?

要拿來炒的粉絲泡水的時間不要太長，以免炒的時候變得太糊爛。可以用冷水或溫水浸泡，雖然用熱水泡比較快軟，但也很容易泡得太軟，不好掌握。

If the vermicelli is meant to be stir-fried, do not soak it for too long or else it will become too mushy. Either cold or warm water may be used for soaking. Though the vermicelli softens faster when soaked in hot water, it becomes overly soft and limp easily.

粉絲炒蛋
stir-fried eggs and mung bean vermicelli

TASTY TIPS

蛋不要炒得太老，因為過後還要和其他材料一起拌炒。
也可撒一點酒增香。

- Make sure that the eggs are not overcooked as they still need to be cooked with the other ingredients at a later stage.
- Add wine for a better aroma.

材料
粉絲	2把
韭菜	3根
蛋	2個
木耳絲	½杯

調味料
醬油	½大匙
高湯或水	1杯
鹽	適量
麻油	數滴

做法

1 粉絲泡軟，略剪短；韭菜切段。

2 蛋加少許鹽打散，再用1大匙油把蛋炒熟，盛出。

3 燒熱1大匙油，淋下醬油和高湯，再放入粉絲和木耳同煮，以小火煮至粉絲入味，加適量鹽調味。

4 放下韭菜拌炒，最後撒下炒好的蛋拌勻，滴下麻油即可。

Ingredients
2 bundles mung bean vermicelli
3 stalks chives
2 eggs
½ cup shredded black fungus

Seasonings
½ tbsp dark soy sauce
1 cup stock or water
dash of salt
few drops of sesame oil

Method

1 Soak vermlcelli until soft and cut it shorter. Cut chives into sections.

2 Beat eggs with a little salt. Using 1 tbsp of oil, scramble eggs until cooked. Dish out.

3 Heat 1 tbsp of oil and add dark soy sauce and stock. Put in the vermicelli and black fungus to cook over low heat until flavourful. Season with salt.

4 Stir in chives and add scrambled eggs to combine well. Drizzle over sesame oil before serving.

用剩的蛋黃該如何保存？
How do you keep leftover egg yolks?

用剩的蛋黃必須倒入一個密封的罐子裡，再放入冰箱，並且在幾天內用完。
Leftover egg yolks must be stored in an airtight container in the refrigerator and used within a few days.

石榴蛋包
egg parcels

TASTY TIPS

- 一般人打蛋是用筷子快速打，這樣會將空氣打入蛋汁中，形成泡沫，煎出來的蛋皮會太過蓬鬆。只要用筷子繞著碗打散蛋汁，讓蛋汁完全混合均勻就可以了。

- It is common practice to beat eggs hard and fast, however this will create too much bubbles in the egg mixture and the result is a thick omelete rather than a smooth thin crepe. Keep the chopsticks in the bowl and use a stirring motion instead to mix the egg until well combined.

材料

雞胸肉	150克
香菇	4朵
金針菇	½把
芹菜	2-3根
蔥末	適量
高湯	2杯

蛋白皮材料

蛋白	4個
太白粉	1茶匙
水	1大匙
鹽	適量

調味料

酒	1茶匙
醬油膏	1茶匙
鹽	¼茶匙
糖、胡椒粉	各適量
太白粉水	適量

做法

1　雞胸肉切成極細的絲，用少許太白粉水拌醃一下；香菇泡軟，切絲；金針菇切除尾端，再切短；芹菜燙軟，撕成細絲。

2　燒熱2大匙油，爆香蔥末，放入雞絲炒散，再放入香菇和金針菇，加入適量鹽炒勻。

3　蛋白打散，加入調勻的太白粉水和鹽，打勻後再過濾一次。鍋中塗一點油，將蛋白煎成6片，每片約4寸大小的圓薄皮。

4　用蛋白皮來包雞絲料，收口用芹菜絲紮好，放入蒸鍋中蒸3-4分鐘後取出。

5　高湯煮滾，加調味料後淋在蒸好的蛋包上即可。

Ingredients

150g chicken breast meat
4 Chinese mushrooms
½ bundle enoki mushrooms
2-3 stalks Chinese celery
some chopped spring onion
2 cups stock

Egg White Crepe Ingredients

4 egg whites
1 tsp potato starch
1 tbsp water
dash of salt

Seasonings

1 tsp wine
1 tsp soy sauce paste
¼ tsp salt
a little sugar, pepper
a little potato starch mixture

Method

1　Cut chicken breast meat into very thin shreds and mix well with potato starch mixture. Soak Chinese mushrooms and cut into shreds. Trim away the roots of the enoki mushrooms and cut them short. Blanch Chinese celery and tear into thin shreds.

2　Heat 2 tbsps of oil and saute chopped spring onion until fragrant, add chicken and fry until separated. Stir In the mushrooms and season with salt.

3　Beat egg whites, add potato starch mixture and salt. Combine well and strain the mixture. Smear the pan with a little oil and cook the egg white mixture into 6 pieces of crepe, each about 4 inches in diameter.

4　Wrap the chicken fillings with the crepes and secure with Chinese celery shreds. Steam for 3-4 minutes. Remove.

5　Bring stock to a boil and add seasonings. Drizzle over the egg parcels to serve.

如何挑選好的鹹蛋？
How do you choose salted eggs?

拿起一個鹹蛋對著光看一下，好的鹹蛋黃是橘紅色的，蛋白則是透明的，如果蛋黃和蛋白呈灰黑色，或混合在一起，就是壞了的。還可以聞一聞，如果有腥臭味也最好別用。

Hold a salted egg up against the light, a good salted egg has bright orangy red yolk and transparent egg white. If both the yolk and the egg white are greyish in colour or mixed together, this means the egg has gone bad. Discard if it smells bad.

鹹蛋炒苦瓜
stir-fried salted eggs with bitter gourd

TASTY TIPS

- 苦瓜先用滾水湯，可以減少苦味；另一個方法是在切好的苦瓜上撒鹽，也可有效減輕苦味。
- *Blanching bitter gourd removes some of its bitterness. Another way is to sprinkle some salt on the bitter gourd slices.*

材料

苦瓜	1條
熟鹹蛋	2個
蝦米	½大匙
大蒜	2粒（切末）
紅辣椒	1條（切片）

調味料

鹽	適量
糖	適量
胡椒粉	適量

做法

1 苦瓜剖開，去除籽及內膜，切成薄片，用滾水燙一下，撈出。

2 鹹蛋取1個蛋白和2個蛋黃，切小塊。

3 蝦米泡軟，切細末。

4 燒熱2大匙油，爆香大蒜和蝦米，加入鹹蛋同炒，炒至起泡。

5 加入苦瓜再炒勻，加適量水及調味料，拌炒均勻，撒下紅辣椒即可。

Ingredients

1 bitter gourd
2 cooked salted eggs
½ tbsp dried shrimps
2 cloves garlic (chopped)
1 red chilli (sliced)

Seasonings

dash of salt
some sugar
dash of pepper

Method

1 Cut open the bitter gourd and remove the seeds and membrane. Cut into thin slices and blanch in boiling water. Remove.

2 Take one salted egg white and 2 salted egg yolks. Cut into small pieces.

3 Soak dried shrimps and chop finely.

4 Heat 2 tbsps of oil and saute garlic and dried shrimps until fragrant. Stir in the salted eggs and fry until foamy.

5 Add the bitter gourd and combine well. Add some water and seasonings. Stir well and scatter over the chilli. Serve.

蒸三色蛋時,如何避免蒸好的蛋黏著模形?

When making this steamed egg dish, how do you prevent the egg from sticking onto the mold?

在模形中先鋪上一層保鮮膜或塗上一層油,就可以很輕易地把蒸好的蛋扣出。
選用鋁箔模形或鐵製的模形,也較容易扣出蒸蛋。

Lay a sheet of cling wrap on the mold or spray it with oil before pouring in the egg mixture. Use an aluminium or metal mold for easy unmolding.

三色蛋
tri colour steamed egg

材料

皮蛋	2個
熟鹹鴨蛋	3個
蛋	4個

調味料

水	4大匙
油	1大匙
鹽	⅓茶匙

做法

1 皮蛋放入水中煮5-6分鐘，取出，剝殼，每個切6小塊。

2 每個鹹蛋切8小塊。

3 蛋加調味料打散，取一半蛋汁加入鹹蛋，倒入方形模形中，入鍋先以中火蒸5分鐘，再改小火蒸10分鐘。

4 皮蛋放入另一半蛋汁中，再倒入模形中，繼續蒸10-12分鐘。

5 取出，稍涼後扣出，切厚片裝盤。

Ingredients

2 century eggs
3 cooked salted eggs
4 eggs

Seasonings

4 tbsps water
1 tbsp oil
⅓ tsp salt

Method

1 Boil century eggs for 5-6 minutes, remove and peel the eggs. Cut each into 6 pieces.

2 Cut each salted egg into 8 pieces.

3 Beat eggs with seasonings, combine half of the egg mixture with salted eggs and pour into a mold. Steam over medium heat for 5 minutes, then lower the heat and steam for a further 10 minutes.

4 Add century eggs to the remaining egg mixture and pour into the same mold. Continue to steam for 10-12 minutes.

5 Remove and leave to cool before unmolding. Cut into thick slices and serve.

如何讓茶碗蒸的配料露出蛋面？
When making chawanmushi, how do you ensure that the toppings stay suspended in the egg custard rather than sinking to the bottom?

可以用兩階段蒸法，預留一些蛋汁，等蛋蒸好，放上魚板和白果，然後倒入預留的蛋汁，再蒸2-3分鐘即可。

If you want the toppings to stay visible above the custard, steam the egg over two times, first steam half of the egg mixture, then place the toppings on top before pouring in more egg mixture to steam for a further 2-3 minutes.

茶碗蒸
chawanmushi

- 打蛋時儘量不打出泡沫，蒸出來的蛋才會美觀。
- 蛋汁用濾網過濾，才能蒸出柔滑細緻的蒸蛋。
- 小茶杯或小碗在放入蛋汁前先預熱，蒸蛋時導熱速度就會比較快。
- Do not beat the eggs too hard until the mixture turns foamy.
- For a smooth texture, strain the egg mixture.
- To allow the egg to cook faster, preheat the small cups or bowls before pouring in the egg mixture.

材料

水	2¼杯
柴魚片	適量
雞胸肉	80克
魚肉	適量
蛋	4個
魚板	適量
白果	適量

調味料

鹽	適量
太白粉	適量
水	½大匙

做法

1 水倒入小鍋中煮滾，熄火，放入柴魚片浸泡，見柴魚片全部沉入水底即撈棄，過濾湯汁，做成柴魚高湯，放至涼。

2 雞肉切片，和魚肉一起加入調味料抓拌均勻，醃製10分鐘，然後放入滾水中燙一下即撈出。

3 蛋加適量鹽打散，加入2倍量的柴魚高湯調勻，將蛋汁過濾到小茶杯或小碗中，每種材料放一些在蛋汁中。

4 包上保鮮膜，放入蒸鍋中，以小火蒸至蛋汁全凝固後取出。

Ingredients

2 ¼ cups water
1 small handful bonito flakes
80g chicken breast meat
some fish meat
4 eggs
some fish cakes
some ginkgo nuts

Seasonings

dash of salt
a little potato starch
½ tbsp water

Method

1 Bring water to a boil, turn off the heat and steep bonito flakes in the water. Discard bonito when it sinks to the bottom of the pot. Strain the dashi and leave to cool.

2 Cut chicken meat into slices and mix well with seasonings together with the fish meat. Set aside for 10 minutes, then scald briefly in boiling water.

3 Beat egg with a dash of salt, add dashi (amount of dashi is double the volume of the eggs) and mix well. Strain mixture into small cups or bowls. Divide the ingredients into the individual cups.

4 Cover with cling wrap and steam over low heat until the egg custard is set.

煎日式蛋捲的秘訣！

What should you take note of when making tamagoyaki?

蛋捲含糖份，所以容易煎焦，因此要注意控製好火候。最好是用長方形的煎鍋，做出來的蛋捲才漂亮。

Tamagoyaki tends to get burnt easily because of its sugar content, thus it is important to watch the heat carefully when frying. Tamagoyaki turns out best when cooked in a rectangular omelette pan.

- 最後煎的時候煎久一點，略呈焦黃，捲起來便會有紋路。
- 在淋下第二層蛋汁之前，要將之前的蛋輕輕挑起，讓第二層的蛋液流入第一層的底部，這樣才比較好捲起來。
- 第二層的蛋汁不要煎太焦，因為之後還有一層蛋汁要捲起。
- *Pan-fry the tamagoyaki well so that you get some grill marks on the egg roll after shaping it on a bamboo mat.*
- *Before pouring in the second layer of egg mixture, lift up the first roll gently to allow the egg mixture to go under the first roll. This will make it easier to roll up another layer.*
- *Make sure that the second layer of egg isn't overcooked since it will continue to cook further after the third layer of egg mixture is added.*

銀魚蛋捲
white baits tamagoyaki

材料
銀魚	½杯
蛋	4個
韭菜	3根

調味料
鹽	適量
味霖	2茶匙
太白粉	1茶匙
水	2茶匙

做法
1 將銀魚洗淨，瀝乾水份，用約1大匙油炒至乾香。

2 蛋加入調味料打散；韭菜切細末。

3 燒熱小平底鍋，塗上一層油，倒下量的蛋汁，攤開煎成蛋餅，見蛋汁將凝固至7分熟時，從鍋邊捲起成筒狀。

4 捲至邊緣時，再倒下量的蛋汁，並撒下銀魚和韭菜末，再邊煎邊捲，捲完後再淋下蛋汁，再捲起。

5 做好蛋捲後改小火，慢慢煎至熟，取出蛋捲，用壽司竹簾捲起定形，切塊排盤。

Ingredients
½ cup white baits
4 eggs
3 stalks chives

Seasonings
dash of salt
2 tsps mirin
1 tsp potato starch
2 tsps water

Method
1 Rinse white baits and drain dry. Using 1 tbsp of oil, fry white baits until they are dry and fragrant.

2 Beat eggs with seasonings. Chop chives.

3 Heat an omelette pan and smear the pan with oil. Pour in ⅓ of the egg mixture and spread out the egg. When it is about 70% cooked, roll up the omelette from one side of the pan.

4 Roll to the opposite edge of the pan, then pour in another ⅓ of the mixture. Scatter white baits and chives on top. Continue to roll up the second layer. Repeat with the remaining egg mixture.

5 Pan-fry the egg roll over low heat until it has cooked through. Remove from the pan and shape tamagoyaki on a bamboo mat. Cut into slices and serve.

蛋餃要如何冷凍保存？
How do you prepare egg dumplings to keep for later use?

自己在家做蛋餃，餡料飽滿，也較衛生，可以一次多做一些冷凍起來。蛋餃做好後要先蒸5-6分鐘至熟，然後才放入冰箱冷凍保存。

Home-made egg dumplings are the best as they have more ingredients and are more substantial. Make more and keep in the refrigerator for later use. You need to steam them for 5-6 minutes until they are cooked, then store in the freezer.

蛋餃燒粉絲
braised egg dumplings with mung bean vermicelli

TASTY TIPS

- 可以用豬絞肉或雞絞肉；豬絞肉的脂肪多，味道較濃郁，肉質也很軟嫩；雞絞肉的脂肪較少，味道較為清淡。

- 要用來包卷材料的蛋皮，記得在蛋液中加入少許太白粉水，以增加蛋皮的彈性。

- Pork or chicken minced meat may be used. Pork has more fats and is more tender and flavourful, chicken is leaner and has a milder flavour.

- When making egg dumplings, add some potato starch mixture in the egg mixture so that the egg crepes can hold the fillings better.

材料

蔥	1根
絞肉	150克
蛋	5個
小白菜	適量
粉絲	2小把

醃料

酒	1茶匙
醬油	1茶匙
鹽	¼茶匙
水	½大匙
太白粉	½茶匙

調味料 1

鹽、太白粉、水 各適量

調味料 2

醬油	2茶匙
鹽	¼茶匙
高湯	160毫升

做法

1 蔥切成蔥花,與絞肉一起再剁一下,放入大碗中,加入醃料拌勻。

2 蛋加適量鹽打散,再加入少許太白粉水,過篩備用。

3 鍋燒熱,改成小火,在鍋子中間塗上少許油,放入1大匙的蛋汁,轉動鍋子,使蛋汁成為橢圓形。

4 趁蛋汁未凝固時,在蛋皮中央放上大匙的肉餡,將蛋皮覆蓋過來,稍微壓住,使蛋皮周圍密合,略煎10秒鐘,反面再煎5秒鐘便可盛出。全部做好備用。

5 小白菜洗淨;粉絲泡軟,剪短一點。

6 燒熱1大匙油,放入小白菜略炒,放入調味料2,將蛋餃放在菜上,以小火煮3-4分鐘,加入泡軟的粉絲,再燒一下即可。

Ingredients

1 stalk spring onion
150g minced meat
5 eggs
some baby bok choy
2 bundles mung bean vermicelli

Marinade

1 tsp wine
1 tsp dark soy sauce
¼ tsp salt
½ tbsp water
½ tsp potato starch

Seasonings 1

some salt, potato starch, water

Seasonings 2

2 tsps dark soy sauce
¼ tsp salt
160ml stock

Method

1 Chop spring onion and mince again with the minced meat. Combine well with marinade.

2 Beat eggs with a little salt, then add some potato starch mixture, strain the egg mixture and set aside.

3 Heat pan and lower the heat, smear a little oil at the centre of the pan and pour in 1 tbsp of egg mixture, swirl the pan to make it into an oval shape.

4 Before the egg has set, place ½ tbsp of minced meat at the centre and fold the egg over to make a dumpling. Gently press the sides to seal. Pan-fry briefly for 10 seconds, then flip over and fry for another 5 seconds. Continue to make more dumplings.

5 Rinse the baby bok choy. Soak mung bean vermicelli and cut it shorter.

6 Heat 1 tbsp of oil and fry baby bok choy briefly, then add seasonings 2, place the egg dumplings on the vegetables and cook over low heat for 3-4 minutes. Add the mung bean vermicelli and cook further until flavourful.

炸蛋酥不求人！
How to make deep-fried crispy eggs?

首先把2個蛋打散；燒熱5-6杯油，約有7-8分熱時改小火，把蛋汁透過漏勺流入油中（圖A）；邊倒蛋汁，邊轉動蛋汁和漏勺（圖B）；倒完之後開中火炸，邊炸邊推動蛋酥（圖C），待顏色快成金黃色時，改成大火，炸5-6秒鐘即撈出（圖D），靜置滴油便可。

Beat 2 eggs. Heat 5-6 cups of oil in a pan. Once the oil is fairly hot, turn down the heat and gradually drizzle in the egg mixture over a strainer (Fig.A). Move the strainer around the pan as you pour (Fig.B). Once all the egg mixture is in the pan, switch to medium heat and push the eggs around the pan (Fig.C). When they are almost golden brown, turn up the heat and cook for a further 5-6 seconds before dishing out (Fig.D). Drain away the oil.

A

B

C

D

TASTY TIPS

- 蛋酥加入白菜中一起滷，可以吸收湯汁的美味。
- 也可用蝦米、干貝或蝦皮來取代扁魚乾。
- Adding deep-fried crispy eggs to the braised cabbage allows the eggs to soak in the flavour of the gravy.
- Dried shrimps, dried scallops and dried shrimp skins may also be used instead of dried sole.

蛋酥白菜滷
braised cabbage with deep-fried crispy eggs

材料
蛋	2 個
扁魚	2-3 小片
大白菜	600克
香菇	3朵
蔥	2根
豬肉絲	100克

調味料
高湯或水	½杯
鹽	½茶匙
烏醋	½大匙
胡椒粉	⅙茶匙
麻油	適量

做法
1 蛋打散，做成蛋酥。

2 將扁魚用溫油、小火，慢慢炸成金黃色，撈起，放涼後剁成細末。

3 白菜切成寬條；香菇泡軟，切條；蔥切段。

4 燒熱2大匙油，放入豬肉、香菇和蔥段炒香，加入白菜同炒至白菜微軟。

5 加入蛋酥和扁魚末略加炒合，倒入高湯或水，燒煮10-12分鐘。

6 加鹽調味，燒至白菜已軟，再加入烏醋、胡椒粉和麻油拌勻即可。

Ingredients
2 eggs
2-3 small slices dried sole (pee hu)
600g Chinese cabbage
3 Chinese mushrooms
2 stalks spring onion
100g shredded pork

Seasonings
½ cup stock or water
½ tsp salt
½ tbsp black vinegar
⅙ tsp pepper
few drops of sesame oil

Method
1 Beat eggs and make them into deep-fried crispy eggs.

2 Deep-fry dried sole in warm oil over low heat until golden brown in colour. Remove and leave to cool, then chop finely.

3 Cut cabbage into thick shreds. Soak Chinese mushrooms and cut into shreds. Cut spring onion into sections.

4 Heat 2 tbsps of oil, fry pork, Chinese mushrooms and spring onion until fragrant. Add cabbage and fry until cabbage is fairly soft.

5 Stir in deep-fried crispy eggs and dried sole. Pour in stock or water and cook for 10-12 minutes.

6 Season with salt and braise until cabbage is soft. Add black vinegar, pepper and sesame oil. Serve.

炒蛋變化吃！
What goes well with scrambled eggs?

許多材料都可以用來搭配炒蛋，例如：韭菜、九層塔、火腿、香腸、培根、洋蔥等。

You can cook scrambled eggs with different ingredients to get a wide variety of tastes. Chives, basil, ham, sausages, bacon and onion can all be added to give the eggs a unique flavour.

炒蛋中加銀芽，可以增加炒蛋的脆口度，吃起來更爽口。

蝦仁用太白粉和鹽醃制，炒後更加爽口。

炒蛋時，油要多一些，火要大一些，才能炒出蛋的香氣。如果喜歡較嫩一點的口感，也可在蛋汁中加一些水，會使蛋炒得更滑嫩一些。

- *Adding bean sprouts to the eggs give the dish a crisp refreshing crunch.*
- *Marinating shrimps with potato starch and salt gives them a succulent crisp texture when cooked.*
- *Use adequate oil and high heat when frying eggs for a better fragrance. Add a little water to the eggs for a more tender texture.*

蝦仁炒蛋
scrambled eggs with shrimps

材料
蝦仁	10只
蛋	5個
銀芽	1杯
蔥花	1大匙

醃料
鹽	⅛茶匙
太白粉	½茶匙

調味料
鹽	½茶匙
胡椒粉	適量

做法
1. 將蝦仁洗淨，擦乾水分，拌上醃料，醃製15分鐘。
2. 蛋加入½茶匙鹽打散；銀芽洗淨，放入蛋汁中備用。
3. 燒熱2大匙油，放入蔥花和蝦仁拌炒，蝦仁一變色隨即倒下蛋汁，翻炒至蛋汁凝固，撒下胡椒粉，盛出。

Ingredients
10 shelled shrimps
5 eggs
1 cup bean sprouts (roots removed)
1 tbsp chopped spring onion

Marinade
⅛ tsp salt
½ tsp potato starch

Seasonings
½ tsp salt
dash of pepper

Method
1. Rinse shrimps and pat dry. Add marinade and set aside for 15 minutes.
2. Beat eggs with ½ tsp of salt. Rinse bean sprouts and add to the eggs. Set aside.
3. Heat 2 tbsps of oil and saute spring onion and shrimps until the shrimps change colour. Pour in the eggs and fry until the eggs have set. Add pepper. Serve.

番茄炒蛋，蛋和番茄的比例要注意！
Why is it important to get the proportion of eggs and tomatoes right when cooking this dish?

如果因為喜歡番茄的味道而多放了番茄，炒出來的蛋就會太酸，湯汁也會太多；
如果番茄放得太少，則味道不足，炒蛋也較乾澀。

The dish will be too sour if you add too many tomatoes and it will also be too wet. On the other hand, it will be too dry if you add more eggs and fewer tomatoes.

TASTY TIPS

- 炒蛋時，要用大火快速攪拌，因為火太小，炒的時間太長，就會把蛋炒得太過老，失去鬆軟的口感。
- 炒番茄時加入一點糖，可以中和番茄的酸味。加糖也可以使炒蛋更加軟嫩。
- Fry eggs briskly over high heat. If the heat is too low, the eggs will take longer to cook through and will become rubbery and tough.
- A little sugar is added to balance the sourness of the tomatoes. It also makes the eggs more tender.

番茄炒蛋
tomato fried eggs

材料
番茄	2個
蛋	4個
蔥	1根（切段）

調味料 1
鹽	¼茶匙

調味料 2
醬青	1大匙
糖	1茶匙
鹽	¼茶匙
水	160毫升

做法
1 番茄的頂端用刀子切割4刀，切成口字形，在尾端切十字刀口，放入開水中燙約1分鐘至番茄皮裂開，撈出泡入冷水中，過一會兒便可將皮剝掉，再切成塊。

2 蛋加入調味料1打散，燒熱2大匙油，將蛋炒至熟，儘量炒成大塊，盛出。

3 另用1大匙油將蔥段爆香，放入番茄炒片刻，加入調味料2，以小火煮3分鐘。

4 放入炒好的蛋，炒勻後再燒1分鐘即可。

Ingredients
2 tomatoes
4 eggs
1 stalk spring onion (cut into sections)

Seasonings 1
¼ tsp salt

Seasonings 2
1 tbsp light soy sauce
1 tsp sugar
¼ tsp salt
160ml water

Method
1 Cut a square at the top of the tomato and a cross at the bottom. Blanch tomato in boiling water for about 1 minute or until the skin separates. Immerse in cold water, then remove and peel the skin before cutting into pieces.

2 Beat eggs with seasonings 1. Heat 2 tbsps of oil and pour in the egg mixture, fry into big chunks of egg. Remove.

3 Heat another 1 tbsp of oil and saute spring onion until fragrant. Stir in tomatoes and fry briefly. Add seasonings 2 and cook over low heat for 3 minutes.

4 Return eggs to the pan and combine well, cook for another 1 minute.

如何避免茶葉蛋的蛋殼脫落？

How to prevent the shells from coming off when cooking these tea eggs?

在蛋殼上輕輕敲出裂痕，可以使蛋入味，但是不要敲得太多，以免蛋殼整個脫落。

Gently tap the eggshell to crack it but do not crack too much so that it still remains intact

茶葉蛋
tea flavoured eggs

TASTY TIPS

- 也可以½杯紅茶葉取代茶包。
- 用滷包來做茶葉蛋也行。
- ½ cup of tea leaves may be used instead of tea bags.
- Pre-packed Chinese stew mix may also be used to make these tea flavoured eggs.

材料
蛋	10個
紅茶包	6袋
八角	1顆

調味料
| 醬油 | 2大匙 |
| 鹽 | 1大匙 |

做法

1　蛋放入鍋中,加冷水,水中加少許鹽或醋,以大火煮滾後改中小火煮約10分鐘。剛開始煮時,要用筷子在水裡轉動蛋。

2　煮好的蛋泡入冷水中,略涼後,用叉子或小茶匙在蛋殼上輕輕敲出裂痕,但不要把蛋殼敲破。

3　鍋中倒入6杯水,加入紅茶包、八角、蛋、醬油和鹽,煮滾後改小火煮1小時以上,關火,浸泡2小時後便可食用。

Ingredients
10 eggs
6 black tea bags
1 star anise

Seasonings
2 tbsps dark soy sauce
1 tbsp salt

Method

1　Place eggs in a pot with cold water (add a little salt or vinegar to the water), bring to a boil over high heat, then boil over medium low heat for about 10 minutes. Using a pair of chopsticks, keep stirring the eggs in the water at the initial stage of boiling.

2　Immerse cooked eggs in cold water. Then use a fork or teaspoon to lightly crack the egg shells, take care not to break the shells too much until they come off.

3　Pour 6 cups of water in a pot, add tea bags, star anise, eggs, dark soy sauce and salt. Bring to a boil, then simmer over low heat for at least an hour. Turn off the heat and let the eggs steep for 2 hours to further develop the colour and flavour.

煮皮蛋粥，要將皮蛋和米抓勻後才煮!
Why do you need to mix century egg with the rice grains before cooking?

抓拌過皮蛋的米較容易煮爛，煮好的粥很滑嫩，而且還有皮蛋的香氣。

Mash and then adding the century egg to the rice grains gives the cooked porridge a smoother texture and a better fragrance too.

皮蛋瘦肉粥
pork and century egg congee

材料

瘦豬肉	100克
皮蛋	2個
米	1杯
高湯	1杯
蔥花	適量

醃料

鹽	¼茶匙
水	1大匙
太白粉	1茶匙

調味料

鹽	適量
胡椒粉	適量

做法

1. 瘦肉切成薄片，加入醃料拌勻，放置30分鐘以上。
2. 皮蛋剝殼，一個切塊，一個不切。
3. 米洗淨放入鍋中，先將一個皮蛋和米一起抓勻，皮蛋抓碎，加入高湯和6杯水，煮滾後改小火煮1小時，煮至粥稀爛。
4. 放入瘦肉和切小塊的皮蛋，以小火再煮5-6分鐘。
5. 加入適量鹽和胡椒粉調味，裝小碗後撒上蔥花。

Ingredients

100g lean pork
2 century eggs
1 cup uncooked rice
1 cup stock
some chopped spring onion

Marinade

¼ tsp salt
1 tbsp water
1 tsp potato starch

Seasonings

dash of salt
dash of pepper

Method

1. Slice pork and mix well with the marinade. Set aside for at least 30 minutes.
2. Peel the century eggs. Cut one egg into pieces.
3. Place rice grains in a pot, crush the other century egg and mix well with the grains. Add stock and 6 cups of water. Bring to a boil, then cook over low heat for 1 hour until the porridge is smooth.
4. Add pork and the century egg pieces. Cook over low heat for a further 5-6 minutes.
5. Season with salt and pepper. Divide into small bowls and scatter chopped spring onion on top.

輕鬆分離蛋黃和蛋白！
How do you separate egg yolks from the whites?

如果有分蛋器就可以輕易分離蛋黃和蛋白，只要把分蛋器架在小碗上，打入雞蛋，蛋白就會沿著縫流入碗中，蛋黃則保留在分蛋器上方。

An egg separator helps separate the yolks from the whites easily. Just place the separator on a small bowl and break an egg into it, the egg white will run down through the slits leaving behind the yolk.

杏汁蛋白
egg whites with apricot kernel juice

材料

南北杏	½杯
蛋白	3個
冰糖	160克

做 法

1 將南北杏泡水2小時，瀝出杏仁，放入果汁機中，加入2杯水打勻，放入布袋中，擠出杏仁汁。

2 杏仁渣與1杯水攪勻，再用果汁機打一遍，擠出杏仁汁，再將兩次的杏仁汁混合煮至沸騰。

3 蛋白打散，倒入煮滾的杏仁汁中，邊倒邊攪打，倒完蛋白後蓋上鍋蓋，燜1分鐘。

4 冰糖加水煮成糖水，隨杏仁蛋白一起食用即可。

Ingredients

½ cup dried apricot kernels
3 egg whites
160g rock sugar

Method

1 Soak dried apricot kernels for 2 hours. Drain and place in a blender with 2 cups of water. Blend well, then squeeze out the juice using a cheesecloth.

2 Combine the leftover kernels crumbs with 1 cup of water and blend again. Squeeze for more juice. Combine it with the juice extracted earlier on and bring to a boil.

3 Beat egg whites and pour into the apricot kernel juice, stirring constantly. Cover with a lid and simmer for 1 minute.

4 Add water to the rock sugar to make a syrup. Serve with the cooked egg whites.

如何讓蛋塔的蛋液烤好後光滑細緻？
How do you get a smooth custard when making egg tarts?

蛋汁加入其他材料拌勻後一定要過濾以去掉雜質，這樣蛋塔餡就會光滑細緻，非常漂亮。
After mixing the egg with the other custard ingredients, it is very important to strain the mixture through a sieve to get a smooth consistency. In this way, the custard will be even and silky smooth.

酥皮蛋塔
egg tarts

- · 蛋汁要儘量減少泡沫，否則烤好的蛋塔餡會有坑洞不漂亮。
- · 糖的份量可以按個人口味酌減。
- · *Make sure that the egg mixture is free from bubbles, or else the custard will not be smooth and even.*
- · *Amount of sugar can be adjusted according to one's taste.*

酥皮料 1

奶油	50克
白油	50克
糖粉	80克
鹽	2克
蛋汁	50克

酥皮料 2

低筋麵粉	200克
泡打粉	⅛茶匙

餡料 1

糖	100克
奶粉	20克
鹽	⅓茶匙
水	200克

餡料 2

蛋	160克
蛋黃	40克

做法

1 將奶油、白油、糖粉和鹽先打發，再分次加入蛋汁，待蛋汁全部吸收後加入酥皮料2，壓拌均勻成團，醒30分鐘後開成長方形的薄面皮，用模形壓出面皮，約20克。

2 將面皮放入模內推開，皮要比模高約0.5-1公分，放入冰箱中醒1小時備用。

3 將餡料1混合均勻，加入餡料2打勻後過濾。

4 將餡灌入模形中約7分滿，蛋塔皮的邊緣刷上少許蛋黃。

5 烤箱先預熱至160C，放入蛋塔，烤約20-25分鐘，至蛋塔餡中心微微晃動即可。

Pastry Ingredients 1
50g butter
50g shortening
80g icing sugar
2g salt
50g beaten egg

Pastry Ingredients 2
200g cake flour
⅛ tsp baking powder

Egg Custard Filling 1
100g sugar
20g milk powder
⅓ tsp salt
200g water

Egg Custard Filling 2
160g egg
40g egg yolk

Method

1 Beat butter, shortening, icing sugar and salt. Gradually add beaten egg, then add pastry ingredients 2. Work into a dough and set aside for 30 minutes. Roll out dough and cut into pieces using a mould, each about 20g.

2 Press dough into tart molds, the dough being ½ -1cm higher than the mold. Leave in the refrigerator for 1 hour.

3 Combine well egg custard filling 1, add filling 2 and whisk well. Strain mixture.

4 Pour mixture into tart shells until about 70% full. Brush the edge of the tart shells with a little egg yolk.

5 Preheat oven to 160ºC, bake egg tarts for about 20-25 minutes until egg custard is set.

為什麼要在放置模形的烤盤內加水？
Why is a water bath necessary when baking an egg pudding?

烤盤中要加熱水，水要深及布丁模形一半的高度，這樣熱水會慢慢的把熱度傳到布丁中，而且熱水使烤箱內有水氣，可以避免布丁的表面乾裂。

Pour hot water into the roasting pan to come half-way up the sides of the ramekins. This prevents the delicate pudding from drying out and allows it to cook evenly.

TASTY TIPS

· 煮焦糖時要注意不要攪動糖，以免糖還原成顆粒狀。成為焦黃色時就要熄火，以免產生焦苦味且太濃稠。

· 鮮奶倒入蛋汁時，關鍵在於不要把蛋汁煮熟了，否則就會有蛋花產生。可以慢慢地把少量的鮮奶倒入蛋汁中，然後再徐徐加入剩餘的鮮奶。

· 在烤盤中倒入熱水時，千萬不可以讓水濺到布丁模形中。

· While making the caramel, do not stir it as the sugar might crystallize and form lumps. Remove from the heat once it caramelizes or else it will become bitter and too thick.

· To avoid heat from the milk from cooking the eggs and thus getting bits of coagulated eggs, slowly add a small amount of milk to the eggs first before combining the remaining milk and the eggs.

· Be careful not to splash water into the ramekins when pouring hot water into the roasting pan.

烤雞蛋布丁
caramel egg pudding

焦糖料

糖	100克
水	2大匙

材料

鮮奶	500毫升
糖	75克
香草精	適量
蛋	4個

做 法

1 小鍋中放入糖和水，以小火煮至糖融化，繼續煮至糖成為焦黃色且濃稠，趁熱倒入模形中，待涼。

2 鮮奶加入糖和香草精煮至糖融化，熄火。

3 蛋打散，將鮮奶倒入蛋汁中，邊倒邊攪打蛋汁，全部加入後，將蛋汁過濾，裝入模形。

4 再把模形放在深烤盤上，烤盤中要加水，水要深及布丁模形一半的高度。

5 烤箱預熱至130-140°C，放入烤約40-50分鐘，至布丁已凝固，用牙籤試一下是否已熟。

6 將模形放入冷水中泡一下，略降溫後即可扣出，或放入冰箱中冷藏後食用。

Caramel

100g sugar
2 tbsps water

Ingredients

500ml fresh milk
75g sugar
a little vanilla essence
4 eggs

Method

1 Boil water and sugar in a small pot over low heat until the sugar melts. Continue to boil until the sugar caramelizes. Pour into ramekins while It Is hot. Leave to cool.

2 Boil milk with sugar and vanilla essence until sugar melts. Turn off the heat.

3 Beat eggs and gradually pour milk into the eggs, stirring constantly. Then strain the mixture and pour into the ramekins.

4 Place ramekins in a deep roasting pan, pour hot water into the pan to come half-way up the sides of the ramekins.

5 Preheat oven to 130°C-140°C, bake for about 40-50 minutes until the puddings are set. Check with a toothpick, it should come out clean.

6 Place ramekins in cold water for a short while and you will be able to easily turn it over onto a serving dish. These may also be chilled before serving.

·怎麼做 最好吃

書　　名/ 怎麼做蛋最好吃

作　　者/ 美食編輯小組企劃

發 行 人/ 程安琪

總 策 劃/ 程顯灝

封面設計/ 洪瑞伯

出 版 者/ 橘子文化事業有限公司

總 代 理/ 三友圖書有限公司

地　　址/ 106台北市安和路2段213號4樓

電　　話/ (02) 2377-4155

傳　　真/ (02) 2377-4355

E-mail /service @sanyau.com.tw

郵政劃撥： 05844889　三友圖書有限公司

總經銷/貿騰發賣股份有限公司

地址/台北縣中和市中正路880號14樓

電話/ (02) 8227-5988

傳真/ (02) 8227-5989

http://www.ju-zi.com.tw
橘子&旗林 網路書店

初版/　2009 年 12 月

定價：新臺幣　149　元

ISBN ： 978-986-6890-63-5 （平裝）

國家圖書館出版預行編目資料

怎麼做蛋最好吃/「橘子文化事業有限公司」
美食編輯小組企劃– –
初版， – –臺北市:橘子文化, 2009.12
面 ； 公分
ISBN 978-986-6890-63-5(平裝)
1. 蛋食譜

427.26　　　　　98019289